Dieter Mende
EEZ Energy, Energy industry, Future energies

Recognizing and questioning fake news and fake videos against the energy transition.
What makes people so susceptible to this?
What can we still use as a guide?

This energy transition is not the first energy transition.

The globally booming energy transition job engine hydrogen is giving the energy markets unprecedented flexibility and dynamism.

If someone tells you that they can explain the energy transition to you within a few minutes, then you should be extremely skeptical.

Thinking at EU/federal/state level and acting locally is not a contradiction, but rather a dynamic energy policy.

Dieter Mende

Dieter Mende
EEZ Energy, Energy industry, Future energies

Recognizing and questioning fake news and fake videos against the energy transition.
What makes people so susceptible to this?
What can we still use as a guide?

This energy transition is not the first energy transition.

The globally booming energy transition job engine hydrogen is giving the energy markets unprecedented flexibility and dynamism.

Impressum

© 2024 **Dieter Mende**, EEZ Energie Energiewirtschaft Zukunftsenergien
www.eez-mende.de

weitere Mitwirkende:

Antje Mende,   LIKES Layout – Impuls – Konzept – Entwurf – Style;
Moderne Medien-, Text- und Bildberatung

Production and publishing: BoD – Books on Demand, Norderstedt

ISBN: 978-3-7597-1371-1

Contents

Prologue

My motivation for creating reports and books is, on the one hand, a passion for highlighting the opportunities and possibilities in the potential grid of the energy transition with hydrogen as an energy carrier and, on the other hand, the ambition to build and expand a hydrogen infrastructure with the advertising of cross-industry service providers, with the identification of sustainable contributions and with the resulting and complementary competencies.

The target group of this book is broadly diversified; the book is aimed on the one hand at readers who are less interested in technology, and on the other hand at schoolchildren and teachers, as well as politicians, entrepreneurs and technicians.
In order for this to succeed, the basis is created at the beginning of the book so that everyone can recognize the starting point of the energy transition.
For this reason, the course of the book is designed in such a way that it does not demand too much basic technical understanding from readers with a lesser affinity for technology, so that the course of the book does not appear boring for entrepreneurs and technicians.

As a result, the first chapter should provide readers with a common basic understanding of the energy transition. The second chapter provides readers with a holistic insight from the sources, the regenerative paths of energy generation, to the sinks, the mobile and stationary applications.

Let yourself be inspired by the fact that change opens up many opportunities, that change is very exciting without exaggeration and that change can trigger a lot of enthusiasm for the future.

Please do not allow yourself to be unsettled by the deliberately created confusion on the part of the lobby against an energy transition, as the lobby against an energy transition deliberately presents incomplete arguments.

The energy transition must be viewed holistically so that the interrelationships that have a significant influence on climate change and environmental impacts can be recognized.

The energy transition must be viewed holistically so that the sound technological expertise and infrastructure know-how can be optimally integrated into the existing energy markets and can also expand the existing energy markets.

At the same time, you will learn about what are certainly the most exciting developments in the modern world with the challenges of today; on the one hand with a view to maintaining the security of energy supply for people, and on the other hand with a view to the many opportunities for future generations.

The energy transition is much more than just the increasing use of renewable energies!

The energy transition is a job engine.

If the energy transition is to succeed and if the agreed climate targets are to be met, there is no alternative to the immediate expansion of renewable energies.

This energy transition is not the first energy transition

With the first energy transition, from burning wood to burning coal, the wood lobby played on people's fears by claiming that, for static reasons, it would not be possible to extract enough volume from the ground to ensure a sufficient energy supply.
The European economy was thus predicted to be doomed.
It was also claimed that coal mining could have an impact on plate tectonics, an impact on the movement of the European tectonic plate, so that volcanism could erupt in Europe.

With the second energy transition, from burning coal to burning natural gas, the coal lobby played on people's fears by claiming that, for technical reasons, it is not possible to maintain a consistently high gas pressure in such a large natural gas network.
Once again, the European economy was predicted to be doomed.

With the current energy transition, the European economy is once again being predicted doom, this time by the pro-fossil fuel lobby for coal and oil products.
It is claimed that renewable energy generation from wind and solar power is too variable and that security of supply cannot be guaranteed.

Once again, the European economy is predicted to be doomed due to the expected dark doldrums, for example when the wind does not blow at night and the sun does not shine. We know that this is not correct.

Currently, on the windiest days, up to 65% of the electrical energy that can be generated in principle has to be curtailed to protect the electrical grids from overload. Only as much electrical energy can be fed into an electrical grid as is simultaneously taken from the electrical grid elsewhere. Short-term, local storage of electrical energy is possible with batteries, while medium-term to long-term storage of electrical energy is possible with the production of hydrogen.

Another example of the energy transition can be seen when looking at mobility. With the mobility turnaround, from horse-drawn carriages to motorized vehicles, the carriage lobby played on people's fears by saying that, on the one hand, the required fuel cannot be available in sufficient quantities and that the required fuel is only available in pharmacies, that, on the other hand, the required fuel is much more expensive than the grass for the horses and that, as a result, goods will become considerably more expensive with a very unreliable distribution over time due to defects in motorized vehicles.

The new technology know-how, infrastructure know-how and energy know-how that is emerging as a result of the energy transition is already recognizable today as a booming global job engine. New and sustainable jobs are being created before other jobs are lost, and considerably more new jobs are being created than old jobs are subsequently lost.

Industrialized countries around the world, as well as the USA and emerging and transition countries such as China, India and Brazil, are making use of our energy expertise.
Small and medium-sized enterprises have taken on an increasingly strong role in the development and production of pioneering energy technologies, as well as providing the corresponding services and service offerings.

The excellent qualification of skilled workers, the human capital, is another key to the success story in Europe.

The integration of German industry into European industry began with the coal and steel industry: the European Coal and Steel Community Treaty (ECSC) was drawn up on April 18, 1951 with Belgium, France, Italy, Luxembourg, the Netherlands and the Federal Republic of Germany, with a term of 50 years.

With its numerous, historically active industries, the Ruhr region in Germany was known as "the land of ten thousand fires". With its focus on electricity generation and consumption in Germany, North Rhine-Westphalia is still the No. 1 energy state today.

With the innovation of the German energy industry, conventional energy sources such as crude oil and natural gas have established themselves on the markets alongside the coal industry since 1951. Energy production was later supplemented by nuclear power generation, which turned out to be the most expensive form of energy production, so that France, for example, nationalized energy production.

The decisive role of a location with a view to energy generation, with a view to an energy region, is probably the ability to generate and supply energy economically, efficiently, ecologically, sustainably and in an environmentally friendly manner.

Both an excellent environment with universities and research institutions, as well as the industrial and technical expertise of the economy, are the innovative basis, with regional economic development as the driving force.

This was true for fossil energy production in the past and is true for renewable energies today.

In view of the range of current energy sources and the increasing need for action to reduce emissions, the energy industry and energy efficiency are decisive platforms for action in the current energy transition.

Resilience, tipping points and the rebound effect

Resilience and tipping points are currently popular topics in the media.

Resilience is the ability of ecosystems to adapt to new conditions in order to preserve them and ensure their continued existence.
Looking at the energy transition, it is clear that the ability to adapt to new conditions has only been successful up to a certain point, and that ecosystems are already clearly exceeding their ability to adapt as a result of climate change.

Yes, it is true that, in terms of evolution, animal and plant species have disappeared from the earth time and again, but in each case up to millions of years ago, so that new species could follow the disappearing species. But what we are currently experiencing, e.g. with the decline of some insect species by up to 75% according to NABU, is the result of what we now know to be the wrong actions of humans. Looking at the course of the earth's evolution, humans have triggered so many changes in just a few years that nature has had no chance to adapt.

Reaching the tipping points at which ecosystems suffer irreparable damage with the expected consequences must be avoided.

The avoidance of irreversible global consequences with the expected catastrophic outcome for life on earth can actually be realized with regional and local "swarm contributions".
Swarm contributions are the sum of all small contributions and all project contributions that are planned and implemented in a city or municipality.
The swarm contributions are the collective multiple with an enormous effect of acceptance among the population, as well as in the involvement of the people and the resulting diversity of initiatives.

The realizable local, regional and supra-regional "swarm contributions" can actually promote educational work among the population with their results; the goal is not curiosity among the population, the goal is sustainable acceptance of the local, regional and supra-regional work.

Presenting the results on social media has become a very valuable multiplier!
However, sharing the valuable results does not trigger any momentum of its own. Maintaining communication about the results is an important basis for further achievement of the ambitious goals.

Tipping points is a term that refers to the Earth's climate system with regard to the energy transition. Whenever the scientific fields of biology, physics and/or chemistry are examined more closely, the complex interrelationships become apparent.

The complex system of the earth's climate is no different. In a figurative sense, tipping points is a term based on the sequence of falling dominoes. If the interrelationships of the earth's climate are seriously altered by human influence, so that a natural sequence of events changes, this irreversibly triggers subsequent reactions.

Examples of tipping points often cited in researchers' publications include the melting of the perpetual ice or the collapse of the influence of the rainforests. The actual tipping points are numerous and can only be recognized when the Earth's climate is examined even more closely. The increase in the earth's temperature is already showing subsequent reactions with the regional drought months, with the regional heavy rainfall, with the clear increase in storm events; and this with increasing intensity.

The successes and results often trigger a rebound effect.

The rebound effect is an effect that reduces the intended target through knock-on effects.

With a view to the energy transition and with the aim of reducing energy consumption in the lighting sector, more energy-efficient lamps than incandescent lamps, for example, have been brought onto the market with LED lamps, both in the private residential sector and in the commercial and industrial sectors.

This has led to a change in people's behavior to the effect that the lighting is not switched off as soon as they leave an area of the home or office, for example, because the more energy-efficient LED lamps supposedly result in much lower energy costs. As a result, the significant increase in lighting time due to people's incorrect behavior has significantly reduced the goal of reducing energy consumption.

Another example of the rebound effect is the development of more energy-efficient combustion engines for vehicles. In the private sector in particular, it is clear to see that quite a few people have developed a level of comfort in their everyday lives that leads to them driving their car even for a few hundred meters, e.g. to the bakery, instead of cycling. As a result, the significant increase in vehicle use and people's misconduct has significantly reduced the goal of reducing environmental emissions.

A significant obstacle to the success of efforts to protect the climate and the environment is people's convenience. This is why local and regional dialogue with people is extremely important with the aim of raising awareness.

Picture:
This is a situation that we are increasingly seeing in agriculture too. The warming of the climate with the effects;
Dieter Mende, EEZ Energy, Energy industry, Future energies

Back to the future

The analogy to the movie "Back to the Future" is not far-fetched, since the know-how of fuel cell technologies and hydrogen production is not new at all! Do you remember the transportation route in the movie?

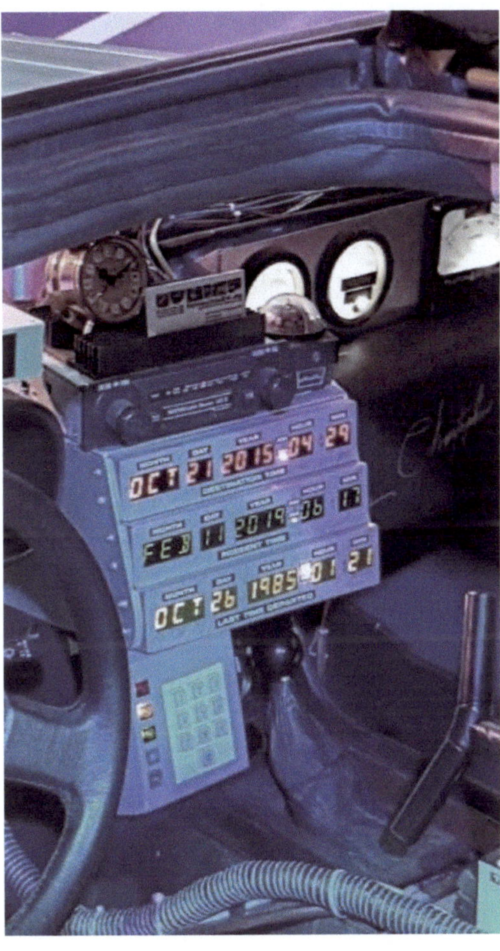

Pictures left and p. 18 above and below:

In the movie "Back to the Future", Marty McFly travels back in time with the help of a time machine, a converted Delorean DMC car, designed in the movie by Dr. Emmett L. Brown.

The vehicle model from the movie was exhibited in February 2019 at the Energy&Water trade fair in Germany/Essen.

Photographed by Dieter Mende, EEZ Energy, Energy Economy, Future Energies

For the energy transition to succeed, we really don't need to invent time travel; the energy transition will succeed with the system integration of the freely available technological know-how of hydrogen technologies and fuel cell technologies.

Hydrogen production with electrolysis, energy recovery with fuel cells and hydrogen transportation have all been known for a few hundred years and are no longer a technical novelty.

In view of the current situation on the energy markets, as well as in view of the current global warming, it MUST become clear that every single kilowatt hour of electricity generated locally from renewable sources contributes to easing the major challenges on the global energy markets, as well as to reducing energy costs, IF, for example, wind turbines are not shut down in the future because the energy grids can no longer absorb the electricity that can basically be generated.

<u>When the wind turbines are shut down because the energy grids can no longer absorb the electricity that can basically be generated, this electrical energy is irrevocably lost. Any other form of energy storage is therefore preferable, even if, in terms of efficiency, a very small proportion of the energy can actually be lost in the conversion processes.</u>

<u>As a result, with regard to the reduction of the electrical energy that can basically be generated, the production of hydrogen is a supplementary added value with the subsequent result that the additional yields of the stored electrical energy have a cost-reducing effect.</u>

The hydrogen can be used for reconversion into electricity in situations where there is not enough renewable electrical energy available, but the hydrogen can also be used directly with a view to optimizing energy efficiency.

In a sustainable energy infrastructure in the energy transition with sector coupling, the hydrogen produced can not only couple the energy products of electricity with heat, with gas products and with fuel products, the hydrogen produced can also couple the energy applications with Power-to-X.

The energy transition is triggering structural and industrial policy changes.

However, this is not due to replacements, but rather to gradual additions and changes.

<u>Renewable energies in the energy mix of the energy transition are increasingly gaining a foreign policy dimension and renewable energies are topical issues in the EU.</u>

The necessity of the energy transition also became clear with the gas crisis in Ukraine at the beginning of 2006 and showed how important renewable energies are in the energy mix for our energy supply security.

<u>Original statement by EU Commission President Ursula von der Leyen, 11.09.2019:</u>
"I want the Green Deal to become Europe's trademark." Ursula von der Leyen tasked her Vice President Frans Timmermans with overseeing the goal of achieving climate neutrality by the middle of the century.

2022 was the year in which the EU had to realize that dependence on energy supplies from countries outside the EU could well lead to serious cuts in security of supply.
Hydrogen energy storage is now also attracting a great deal of attention from those who have not previously been involved in the energy transition.

Initial answers to questions about the potential of the energy transition, current technical possibilities, research and development, efforts to create a sustainable energy infrastructure, project developments and expectations regarding the timeline for the introduction of new technologies can be found in the regional dialogues in the energy transition potential grid.

<u>Regions have the opportunity to create sustainable and high-quality jobs, not just at the end of the energy transition, both in rural and urban parts of the country.</u>

The new markets follow the potential for expansion and are also of extraordinary interest to industry and service providers, trades and commerce.
The foreseeable penetration of various markets by future energies and technologies, including hydrogen as an energy source and fuel cells as an energy converter, will open up opportunities for traditional production processes; established industrial structures will experience an innovative boost.

Whenever Europe has experienced a turnaround in energy use, numerous companies from different sectors have come together to form strategic alliances. The basis of an economy such as the one we have in Europe is the provision of energy.

A central question of what the energy supply will look like in the future has also come to the attention of those who have not yet dealt with the issue of the energy transition, or who have not wanted to deal with the issue of the energy transition for reasons of their lobbying.

The energy transition also leads to mobility and the regional level is particularly important as it can have a direct influence on transport-related emissions by shaping the infrastructure and controlling the behavior of private and commercial transport users.
Regions act within the framework of federal and state policy requirements through knowledge of local and regional needs; close to the citizens, fast and also cost-optimized. The energy turnaround has been the subject of regional focus for many years.

For example, the German region of Emscher-Lippe is no longer just a model region in North Rhine-Westphalia NRW; it can already present results for the ambitious goals.

Both the energy supply companies and the local public transport companies are important partners in the regional dialogs.

Municipal companies have a role model function; they can show citizens by practical example that it is possible to decouple CO2 emissions and transport services.

Only if this can be experienced can the energy transition also be accepted by citizens with regard to transport and mobility. The regions have recognized what will move us in the future.

<u>The regional model can be committed to sustainable mobility and should go hand in hand with protection against traffic noise and air pollutants.</u>

The mission statement includes the partnership-based design of traffic flows involving all means of individual and public transportation. This leads to new goals and measures.

Efficient technologies should also make road traffic more sustainable. The outstanding importance of electric vehicles for urban regions is undisputed. The use of hydrogen and fuel cell technologies with increased CO2 efficiency is also supported and accompanied; the energy supply companies can expand their business areas with a view to wind power electrolysis; in addition to the potential of power-to-gas, methanation and mobility.

Mobility is not only a basic human need, in the modern world it forms the basis for many professional and private activities of citizens, increasingly with reference to renewable energy sources. Local public transport can also offer solutions for reducing and avoiding emissions.
This requires the continuous improvement of all services such as convenience, safety, punctuality, network density and information.

Mobility is seen as a fundamental right. This positive perception of citizens in Europe also means that energy supply companies are close to the volume markets, with very high added value.

<u>For example, Germany is at the center of the major European goods flows.</u>
<u>The proximity to the volume markets is an important pillar for the positioning of the countries in the potential grid of the energy transition in Europe.</u>

The EU ministries and the ministries of the European countries have recognized that hydrogen is an important storage medium for excess capacities of electrical energy generated, for example, by wind or solar power.
The regional approaches show which complementary technical tangents make hydrogen a suitable energy storage medium. Regional approaches can complement each other and generate a hydrogen agenda in the countries.

At this point, it should be emphasized once again:
<u>Thinking at EU/federal/state level and acting locally is not a contradiction, but rather a dynamic energy policy.</u>

Picture:
The hydrogen user center h2herten in front of the historic Ewald colliery;
Photographed and panorama created by Dieter Mende,
EEZ Energy, Energy Economy, Future Energies

Structural change, regardless of the background, is a huge challenge, but also a huge opportunity for the municipalities and the regions at the same time. The Ewald colliery in Herten worked its last production shift on March 28, 2000 and was finally shut down in 2001. The transition from the old, renowned energies to the new, increasingly regenerative energy mix is also a reflection of the booming global job engine of the energy transition.

The numerous international guests make this clear.

The concept of Ewald as a location for the future has been followed by the implementation of the previously formulated regional goals, creating a unique selling point for the municipality and the region:

h2herten has already attracted a great deal of attention in the EU and also worldwide since the turn of the millennium.

<u>The figures published at the end of 2018 clearly show how different the energy infrastructure in the municipalities and regions of Europe can and must be.</u>
<u>Here using the example of the German state of North Rhine-Westphalia:</u>

+ At the end of December 2015, the population of North Rhine-Westphalia was 17.68 million,
+ At the end of June 2022, the population of North Rhine-Westphalia was 18.07 million people.
+ This corresponds to an increase of 2.16% of the population in North Rhine-Westphalia.
+ At the end of June 2022, the population of the Ruhr area was 5.13 million people.

The structural requirements of urban centers are often contrary to the structural requirements in rural areas.

While at the end of June 2022 in the four largest cities in North Rhine-Westphalia:

+	in Essen		583.153 People,
+	in Dortmund		592.900 People,
+	in Düsseldorf		625.581 People and
+	in Köln		1.081.167 People lived,

lived in the smallest municipalities in North Rhine-Westphalia at the end of June 2022:

+	in Heimbach		4.345 People,
+	in Dahlem		4.395 People.

Decentralized energy generation supports both rural and urban energy infrastructure. Regional pioneers attract a lot of attention in the global market.

Pursuing regional goals with the formation of clusters achieves entrepreneurial success, achieves competitive advantages with specialization also for the companies, achieves the announced intentions with the energy transition with the individual offers in the market.

Regional successes are not achieved by copying existing, neighboring regional successes; on the contrary, these copies weaken the regions concerned through unnecessary competition with each other.
Regional successes can be found in complementary contributions to each other in a country's potential grid. Cooperating regions can complement and strengthen each other both in neighborly cooperation and in international cooperation.

The energy transition has an enormously broad spectrum and offers the regions numerous opportunities for growth. This is confirmed by studies by the BMWi Federal Ministry for Economic Affairs and Energy on the energy transition and its impact on investment, growth and employment.

WiFö Wirtschaftsförderung, for example WiFö in the Emscher-Lippe region, has been able to convincingly demonstrate to the business community that hydrogen and fuel cell technologies are wanted regionally. This requires well-maintained regional information and demonstration work.

Only if cluster management is positively recognizable can doubts about investing in this location be dispelled. The aim is to bundle the opportunities in the technology field of fuel cells and hydrogen for the sustainable creation of future-oriented jobs. The image of a region in the field of fuel cell and hydrogen technologies is assessed on the basis of its networking activities. The development of a region can therefore only be successfully managed if the entire industrial and scientific environment of Europe is included in the project work.

Regional fields of action are identified potentials for success with sustainability. Along the value chain from regenerative hydrogen production to the applications, including the requirements of the hydrogen infrastructure, the niche products and special applications must be positioned sustainably in the early markets.

In addition, highlighting expertise in terms of services and support is of great importance in order to counteract initial skepticism towards a new technology application with correspondingly positive applications.

Both the technology and market potential as well as the growth expectations of the markets are not sustainable without location integration and integration into the European potential grid.

With reference to the initial question about potentials, current technical possibilities, research and development, infrastructure efforts, project developments and expectations on the timeline until the introduction of new technologies: Both the potential links and the formation of competence networks with the economy and the markets enable innovative, regional steering groups through their bundling.

The energy transition, climate change, digitalization and other future fields have already reached numerous international markets globally with technology and infrastructure expertise.
Energy generation is focusing on an energy mix with wind, solar, GEO, BIO and hydro, for example, which is increasingly regenerative.

The decisive factor here is how constructively communication is promoted and how intensively partners are recruited by demonstrating clearly communicated regional goals.

The inclusion of regional cooperation and networked expertise creates suitable platforms for action.

Against the backdrop of international developments in the potential of hydrogen as an energy source, regional opportunities lie in participating in a sustainable infrastructure in the energy transition, which includes the consistent expansion of renewable energy production.

Picture:
KlimaExpo.NRW exhibition during Energy & Water 2015 in Essen;
Dieter Mende, EEZ Energy, Energy Economy, Future Energies

The Emscher-Lippe region is one of the most densely populated regions in the Federal Republic of Germany and forms the northern part of the Ruhr area; it is named after the rivers Emscher and Lippe.

With almost one million inhabitants, the Emscher-Lippe energy region is historically known both as an innovative chemical site and as a former mining and coal mining location, far beyond the borders of the EU.
The structural change in the region is far from complete with the closure of the last coal mine at the end of 2018, but is entering a second phase with the coal phase-out in the energy sector. However, this also offers a real opportunity to innovatively reshape Emscher-Lippe as a business location with its energy-rich industries and to decouple the region from the use of fossil fuels in terms of both energy and materials.

The Emscher-Lippe region comprises the two independent cities of Gelsenkirchen and Bottrop on the one hand and the district of Recklinghausen with its 10 towns on the other. With the exception of the Hanover region, the district of Recklinghausen is the most populous district in Germany.

The following list shows the population figures for the Emscher-Lippe region. December 2018 was deliberately chosen to show the population at the final end of coal mining.

993,298 inhabitants are spread across twelve cities.

993.298 inhabitants are spread across these cities:

+ the independent city of **Gelsenkirchen:**
  260.654 People;

+ the independent city of **Bottrop:**
  117.383 People;

+ the district of **Recklinghausen:**
  615.261 People;

+ the district town **Recklinghausen:**
  112.267 People;

+ the district town **Marl:**
  83.941 People;

+ the district town **Gladbeck:**
  75.687 People;

+ the district town **Dorsten:**
  74.736 People;

+ the district town **Castrop-Rauxel:**
  73.425 People;

+ the district town **Herten:**
  61.791 People;

+ the district town **Haltern am See:**
  38.013 People;

+ the district town **Datteln:**
  34.614 People;

+ the district town **Oer-Erkenschwick:**
  31.442 People;

+ the district town **Waltrop:**
  29.345 People.

A multitude of good ideas alone does not by itself trigger a sustainable infrastructure in the energy transition.
The combination of company-oriented implementation of these ideas with regional opportunities and possibilities leads to innovation-oriented, continuous structural change.

The end of coal-fired power plants, the shrinking of the steel industry and the closure of coal mines mark the start of the third regional transformation stage. It is now important that the energy-intensive industries (chemicals, metal processing, glass production, etc.) are kept in the region despite the phase-out of fossil fuels. The challenges of structural change are seen as an opportunity and a call for constant renewal. The aim of regional economic development is to bundle opportunities in the technology field of energy transition in order to create sustainable, future-oriented jobs.

Bridging technologies and future technologies complement each other and optimize the market ramp-up of "green hydrogen" with a view to energy supply security, environmental compatibility, affordable energy offers and new energy products. New energy products are being created with Power-to-X, sector coupling of electricity, gas products, fuel products and heat.

The image of an energy region in the fields of fuel cell technology and hydrogen technology is assessed on the basis of its networking activities.

The development of a region can therefore only be successfully managed if the entire industrial and scientific environment in Europe is included in the project work; Germany is a role model in this respect and many European countries are already following suit.

This is not a natural process, but a cross-industry network management in which the municipal contributions flow into regional project management.

The different regional starting points often show the incomplete value chains from the sources (renewable energy generation) to the sinks (their mobile and stationary applications). However, the regional starting points are completed by the value chains through cooperation.

If the energy transition succeeds in the Emscher-Lippe region, which is characterized by energy-intensive basic industries, then the energy transition can succeed everywhere in Europe.

The players cooperating in the Emscher-Lippe region in the field of energy transition combine different disciplines with the aim of implementing the best possible solutions.

These are very promising prospects for all dialog partners; in the state of NRW, in Germany, in the EU and also far beyond European borders.

The hydrogen coordination installed in the Emscher-Lippe region bundles these solutions and sensitizes companies and society to the topic of the energy transition with hydrogen.

The change management of the regional economic development agency WiN Emscher-Lippe GmbH and the ambitious commitments take a particularly close look at the focal points:

+ energy,
+ the chemical industry
+ the circular economy,
+ metal processing and other industrial focal points,
+ the healthcare industry
+ agriculture,
+ local recreation,
+ nature conservation,
+ sustainability,
+ the challenges of climate change,

in the former coal and mining region.

The basic prerequisite for the sustainable development of the economy and society is environmentally friendly, or at least environmentally neutral, socially acceptable and economically efficient energy technologies.

The primary goal is therefore to recognize technical, economic and social developments at an early stage in order to trigger the development of innovative technologies.

Picture:
Dieter Kwapis, Manager of the hydrogen user center h2herten and the ZZH Zukunfts-Zentrum-Herten: Stand support at the joint stand during the Hannover Messe 2019. In the background the joint appearance:
ZBT Center for Fuel Cell Technology Duisburg, HyCologne and the h2-netzwerk-ruhr at the stand of the former Energy Agency of North Rhine-Westphalia;
Dieter Mende, EEZ Energy, Energy industry, Future energies

A statement that is currently often quoted in the media is: "The energy transition does not come for free."

However, if you look at the quantifiable damage caused by climate change, you can see that the energy transition is an investment that will save costs in the medium term.
The British market research company Economist Intelligence Unit (EIU) puts the climate crisis and its impact on the global economy at almost eight trillion dollars (approx. 7,200 billion euros) by the middle of the century.
The implementation of the energy transition in the exemplary model region of Emscher-Lippe can sustainably generate a leading regional position in Europe as a job engine.

This is not about decarbonizing the economy, but rather about defossilizing production processes and cyclizing the value chains of carbon as far as possible, which is too valuable as a pure energy source in the long term.

The energy technologies that are important for the success of the energy transition, including hydrogen energy storage, cannot yet be operated economically as the value chains are not yet complete. The energy technologies that are important for the success of the energy transition, including hydrogen energy storage, require the necessary adjustments to legal regulations.

Legal regulations that ensure that regeneratively produced hydrogen is at least not placed in a worse position than fossil fuels.

<u>The energy transition is also the move away from today's fossil-based energies, tied up in coal, tied up in oil products and tied up in natural gas, towards an increasingly regenerative energy economy with a focus on the generation of electrical energy.</u>
<u>This means that there is a lack of important storage facilities for renewable electrical energy.</u>

The previous chapter demonstrated this:
Fields of action are emerging that tap into the numerous opportunities for shaping the future and with which the regions can attract sustainable jobs. Regional models also arise from the pooling of municipal competencies.

Regional success, as has been shown time and time again, lies in municipal cooperation; the parochial thinking of the municipalities, the communities, rather prevents regional developments.

Research into and integration of bridging technologies and future technologies must take place now, i.e. before "green" hydrogen is sufficiently available, in order to achieve the necessary market maturity so that the climate targets set by Europe can be achieved.
To this end, there are various applications for funding from local authorities, business development agencies, networks and universities, all of which are interlinked.

A successful regional energy strategy can be based on the following points:

+ Relieving the grid by absorbing excess capacity generated from renewable sources,
+ decoupling growth from resource consumption,
+ Independence from fossil fuels,
+ Coupling of value chains: H2, O2, CO2 offer themselves,
+ Material utilization possibilities for CO2 reduction,
+ recycling of carbon,
+ Joint mobility and commuter strategies,
+ System integration of bridging technologies,
+ System integration of future technologies,
+ Integration of innovative small and medium-sized enterprises,
+ Creating acceptance among the population,
+ job security and expansion of the professional future.

Successful regional action paths can be:
+   mobility,
+   local public transportation,
+   freight traffic,
+   private applications,
+   commercial applications,
+   industrial applications,
+   leisure activities with applications,
+   cultural events with applications,
+   environmental protection and climate-relevant target agreements,
+   construction and planning of relevant target agreements,
+   infrastructure and sustainability of relevant target agreements,
+   Accompanying economic development: research, development and qualification,
+   Neighborhood development residential,
+   Neighborhood development commercial,
+   Neighborhood development industry.

Attention should be paid to the successful regional action paths:
+   the material flows: Power-to-X,
+   sector coupling: electrical energy with heat, gas and fuel products,
+   environmental compatibility, reducing the impact of climate change.

Regional integration of economic development:
Regional cluster management:

Regional cluster management, which should result in the expansion of jobs, is not a natural process.
Cross-sector network management proactively implements the formulated objectives at project and working levels:
+   the schools and universities,
+   the educational institutions with the additional qualification,
+   the companies with sustainable orientation,
+   the trade unions with accompanying measures,
+   Associations and organizations with a future-oriented approach,
+   politics with accompanying measures.

Once again, the Emscher Lippe region is a good example:
The Emscher-Lippe region was already an active project partner in the EU project HYCHAIN MINI-TRANS - future prospects with hydrogen from 2006 to 2011

"HYCHAIN MINI-TRANS" was an EU flagship project to demonstrate future-oriented applications of hydrogen in the transportation sector, taking into account the entire value chain. The project was successfully completed in June 2011. In the hydrogen user center h2herten alone, many hydrogen-related jobs have since been created.

Fake news, fake videos ... There is an alarming amount of all this on the internet, including on the energy transition. But what can we still use as a guide?

My impulse: the results of serious networking with the confirmation of numerous complementary players.

With a view to the energy transition, with a view to Power-to-X, with a view to hydrogen linking the sectors of electricity, heat, gas products and fuel products, these two networks have been successfully complementing each other in NRW since 2008:
+   h2-netzwerk-ruhr for the Ruhr region,
+   HyCologne for the Rhineland.

We support these developments very successfully; locally, regionally, nationally and internationally with the hydrogen user center h2herten.

Would ideas from Jeff Bezos, Steve Jobs, Larry Page, Sergei Brin, Bill Gates, Mark Zuckerberg and/or Elon Musk currently have a chance in Europe, in Germany?

Probably not from a precise point of view, partly due to the current nagging mentality in Europe, which often looks first at why something supposedly won't work rather than promoting interdisciplinary action.

Economic growth Europe: Today's founders are no longer founding families like the German founding families Krupp, Siemens and others; today's founders call themselves start-ups and are often rather unknown in terms of personnel.

For example, the hybrid technology developed in Germany was completely misjudged in Germany and not integrated into the systems. Only a few years later, almost far too late, after the markets in Asia presented us with mobile and stationary systems with hybrid technology, did the system integration of hybrid technology also take place in Germany.

Does anyone really believe that the incandescent lamp developed by Thomas Alva Edison was the result right from the start?

The steam engine was given a decisive improvement in 1769 by James Watt, on which he was granted a patent.

The German government has presented an update of the National Hydrogen Strategy. Immediately, numerous nagging voices appeared, e.g. also on LinkedIn, commenting on why the National Hydrogen Strategy is unlikely to work. It is not the task of the federal government to optimize an economic sector, such as the energy markets, for example; that is the task of the industries, the companies, the service providers and also the users, but also the municipalities, the regions and the federal states.

The National Hydrogen Strategy is a politically provided basis with which the most diverse markets in the energy industry can be optimized.
With Power-to-X, the energy markets will gain unprecedented flexibility and dynamism through hydrogen, which links the sectors of electricity, heat, gas products and fuel products.

Plant engineering, digitalization, robotics, communication technologies and many more are growing at these tangents. The energy transition job engine is also booming in Europe! Would you like to be part of the energy transition job engine and the global economic growth that is taking off, but are not yet aware of your opportunities and possibilities?

No problem. The companies located in the hydrogen user center h2herten have international expertise and will be happy to show you your opportunities and possibilities after identifying them; they will also be happy to shape them together with you.

At this point, to close the thematic circle with the course of the book:
Yes, the expansion of the electricity grids has not been done sufficiently with a view to the energy transition, because the energy transition is also the path away from material-based energies, bound in coal, bound in oil products, bound in gas products, towards an increasingly regeneratively generated electrical energy.

But yes, the electrical energy grids do not work like a sponge in the electrical energy infrastructure without sufficient energy storage.

Even if our electrical grid were ten times, or even a hundred times, as large as it is now, only as much electrical energy can be fed into the electrical grid as is taken from the grid elsewhere at the same time.

So yes, without the storage facilities for electrical energy, already in the gigawatt range, we are left with the curtailment of renewable energy generation, which means that the electrical energy that can be generated in principle is irrevocably lost when the wind turbines, for example, are curtailed.

Yes, the urgent challenges of the energy transition lie in the complete consideration, but yes, there are also numerous opportunities and possibilities for optimization.
Whenever the energy transition is considered in its entirety, starting with energy generation and extending to mobile, portable and stationary applications, hydrogen becomes apparent at numerous tangents.

The energy transition is a job engine! Wind, solar, geo, bio, hydro ... energy generation is increasingly regenerative and intelligent energy management, along with energy storage, plant construction, engineering, grid dynamization and global market technology know-how, are the triggers for the enormously growing job market.

The argument put forward by supporters of current conventional energy generation that the energy transition is destroying jobs does not stand up to a holistic view. Yes, it is true that one job in conventional energy generation has, on average, generated four more jobs in terms of the extraction of coal, oil and natural gas, plant construction and the development of materials and components.

However, it is also true that one job in renewable energy generation generates an average of four (+) additional jobs in terms of plant construction, materials development and component development, and also new, in terms of decentralized energy generation and the resulting smart grids. Smart and digital are no longer new terms in the energy industry and are creating many new and additional jobs.

In renewable energy generation, material development and component development is no less demanding and has generated the (+) of additional jobs in the average evaluation of jobs that follow a job in renewable energy generation.
The examples of this are numerous and no less diverse than renewable energy production itself.

Take Germany, for example: the figures, data and facts available on the internet show that wind energy generation already outstripped energy generation from nuclear power plants in 2015.

This positive development can also be followed by the maximum added value of renewable energy generation with suitable energy storage systems.

Hydrogen is a climate-neutral and at the same time environmentally friendly energy storage and energy carrier for renewable energies; power-to-gas has become the focus of both global energy suppliers and regional energy suppliers operating in networks.

However, power-to-gas is not only innovative as an energy storage system and energy carrier.
The numerous innovative developments along the potential grid of the energy transition show further innovative developments at many technical tangents.

Renewable energy generation and mobility are already complementing each other and together are leading to a sustainable infrastructure in the energy transition.

So it is no wonder that both the pro-fossil fuel lobby and the pro-nuclear power plant lobby are currently questioning hydrogen so intensively again with deliberately incomplete or even deliberately false reports.
How else can the pro-fossil fuel lobby and the pro-nuclear lobby react other than with deliberately incomplete or even deliberately false reports?

Because with a complete view of the energy markets, the tangents to hydrogen are unmistakably numerous.

<u>The native peoples have a foresight that is often pushed into the background in the so-called modern world and its fast pace of life.</u>

A Native American phrase from the time of the formation of the American colonies is once again becoming a warning in view of the energy transition and the reduction of the effects of climate change:

<u>We do not inherit the earth from our ancestors, we borrow the earth from our children, from our grandchildren and from future generations.</u>

It is not uncommon for what is right and important to conflict with interests, especially when it is also about making money, when it is about ensuring that the current business areas receive the maximum possible added value.

In reference works, the term lobbying is explained as meaning that politics, for example, is subject to systematic and ongoing influence by commercial enterprises, that social and/or societal groups want to influence political decision-makers, that influence is usually to be exerted through personal contact or via the mass media, with a view to shaping people's opinions.

Lobby Control and deputies watch are organizations that critically and clearly point out that financially strong business associations can exert a great deal of power over politics, that financially strong business associations can even manipulate political decisions, while less influential lobby groups have a much harder time attracting attention with regard to other interests.
More transparency, e.g. with regard to the development of legislative proposals, is the focus of the organizations.

It is well known that there is a lobby register or transparency register in the USA and the EU.
In Germany, there has been a lobby register or transparency register since 01.01.2022.
Around 5,000 associations, companies and organizations are listed in the register.
Corruption experts and politicians are calling for stricter regulations with regard to lobbyists in the EU and in state ministries.

Examples from Germany:
Do you remember the lobbying scandal during the corona pandemic, known as the "mask affair" involving members of the Bundestag Nikolas Löbel (CDU) and Georg Nüßlein (CSU), as well as the Bavarian member of parliament and former Bavarian Minister of Justice Alfred Sauter (CSU)?

The German magazine Spiegel had reported that the three gentlemen had personally enriched themselves by looking at the procurement of face masks.

Do you remember the case in 2020 in which MP Philipp Amthor (CDU) caused a stir by allegedly lobbying for an American start-up company and receiving share options in return, as well as a directorship?

Such affairs were also a trigger for the grand coalition in Germany to agree on the introduction of a binding lobby register.
Political work cannot succeed without lobbying. This may seem strange at first, but a closer look reveals that, in principle, lobbying supports political work in the EU and in federal and state ministries.
Only when the various interest groups at the political levels actually have all the necessary information at their disposal can they weigh up the content with a view to preparing legislative texts, for example. Not all politicians can have the same level of specialist knowledge, especially given the large number of specialist areas.

The analogy with a view to the courts:
If judges are asked to pronounce justice in court on a subject in which they are not entirely sure of the content, then judges have the option of commissioning an expert opinion from those who they know are familiar with the subject.

The approach in politics is similar. Politicians, for example, are not necessarily tech-savvy when it comes to the energy transition. As a result, politicians have the option of either commissioning a study or setting up a specialist group.

People who can be assumed to have specialist expertise are invited to the expert group.

The lobby groups have the relevant expertise in the specialist areas. In a democracy, not only must there be freedom of opinion, but also, as with regard to the energy transition, there must be openness to technology without discrimination, provided that this technology does not contradict existing decisions, such as the coal phase-out. Issues can only be decided without discrimination if the entire topic is considered from a wide range of options and with a wide range of potentials; this is very important for the political decision-making process. That is why, before laws are passed, consultation with the associations, consultation with industry, is indeed a political requirement.

One example that illustrates the conflicting lobby interests is the decision-making process that was restarted at the end of 2023 with regard to glyphosate in agriculture. The interest of maximizing profits in agriculture is in conflict with the interest of environmental protection.

Another example has been seen mainly in England with the disease of cows with mad cow disease. It was only after the first cases of illness in humans pointed out that mad cow disease could be transmitted from cows to humans through industrial products such as galantine that the pro-cattle breeding lobby was no longer able to prevail over the pro-healthcare lobby.

A suitable example of conflicts of interest is the German power generation company RWE. RWE elected Jürgen Grossmann as CEO in 2007 with the task of making RWE fit for the future; the advertising slogan was "voRWEg gehen"; the meaning of "voRWEg gehen" is go ahead.

Jürgen Grossmann has always been very innovative and has recognized hydrogen as a pillar of a sustainable energy infrastructure.

In 2010, RWE was the main sponsor of the World Hydrogen Congress in Essen; we were a branch of the World Hydrogen Congress 2010 with the hydrogen user center h2herten and we demonstrated to the congress participants how the sustainable energy infrastructure with hydrogen works; starting with renewable energy generation, through energy storage, to the provision of energy from electricity and hydrogen.

It was the shareholders of RWE who had forbidden RWE CEO Jürgen Grossmann to include hydrogen in the sustainable business areas, because the energy carrier hydrogen competes with the business areas of energy generation with coal-fired power plants and nuclear power plants. It was the shareholders of RWE who imposed the statement on RWE CEO Jürgen Grossmann that RWE stands for coal-fired and nuclear power.

On March 11, 2011, the Fukushima nuclear power plant disaster occurred in Japan, which led to the political decision to phase out nuclear power on the grounds that if Japan's world-leading nuclear technology expertise is not able to prevent such a disaster, then neither can we.
As a person, Jürgen Grossmann would have been exactly the right CEO for RWE, but due to the statement imposed on him by RWE shareholders that RWE stands for coal-fired power and nuclear power, Jürgen Grossmann would no longer have been credible to outsiders with a view to re-election as CEO.

Now, in 2023, with the ramp-up of the hydrogen industry, RWE would be glad not to have made this statement. If RWE can now imagine getting out of energy generation with lignite sooner than the EU stipulates, it is because RWE shareholders have realized that shares that are tangential to the energy transition with hydrogen sell better than shares that are tangential to lignite.

Conclusion: it is not necessarily the managing directors in the companies with AG Aktiengesellschaft in their name, it is the capital, it is the shareholders who can stand in the way of innovation with the aim of maximizing the value creation of the current business areas.

Fake news, fake videos ...
What makes people so susceptible to them?

Don't worry, the course of the book won't go too deeply into psychology, but it really does require an honest approach to the background and the triggers when looking at the prospects of the energy transition and the active lobby for fossil fuels against the energy transition.

Your perceptions influence your decisions and if you surround yourself with negative-minded people, you will not receive any positive impulses.
Flipping the switch, getting out of the sum of negative impulses is mainly up to you, your willingness to question.

As a book author, I am and will continue to be active in the area in which I am very well versed: the energy transition.
At this point, the book is a heartfelt invitation to let go of the hedgehog attitude towards the energy transition and enter into a joint energy transition dialog.

In order to recognize the opportunities and possibilities of the energy transition, we are aware that what is important and what is right always requires honest and open dialogue. With the extended view beyond the Johari window, I will show you more concretely where the connections are and how you can best classify them.

Many of you, dear readers, will probably not have heard the term Johari window before.
However, the way people interact with each other creates precisely this image.

Facebook, Instagram, X (previously Twitter) and YouTube are examples of popular social media that are not only used by stars from the TV and music industry, but also by private users with their own online profiles.

Take a look at the Johari window sketch on the following page 56:
Social media works excellently with acting and reacting:
+   Thus, in the top left of the sketch is the "Public Person", the profile of the participants in social media.
+   "My secret" is the pool of information from which the profile is continuously fed.
+   "Blind spot" is the area of those who react; this can be likes, but it can also mean shitstorms.
+   It gets more exciting for everyone with "Unknown", because this area has a lot of potential.

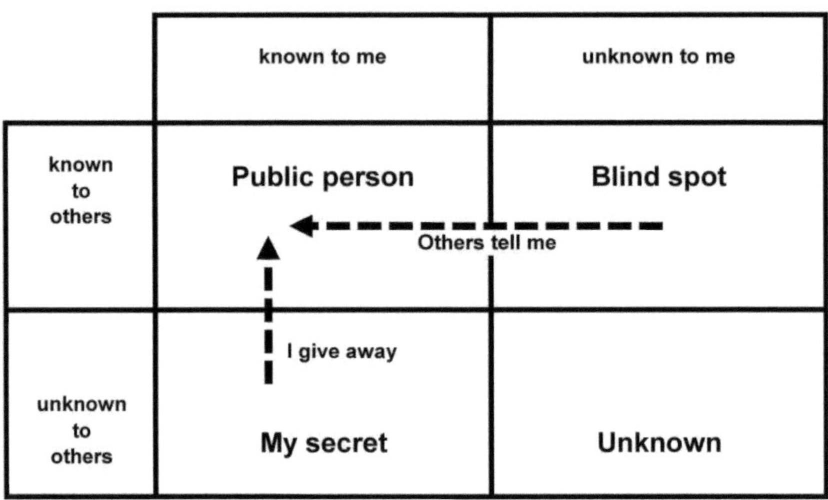

Sketch:
In 1955, the American social psychologists Joseph Luft and Harry Ingham
Harry Ingham introduced the Johari Window.
The first names of the social psychologists, Joseph and Harry, led to the naming
of the Johari Window.

Before we turn our attention to looking beyond the Johari Window, I will outline a methodology that you can use to apply the theme of the Johari Window either from the passive situation, e.g. for obtaining information or even for breaking free from the hedgehog attitude.

The Johari window can and should trigger a dynamic, regardless of whether it is with an individual person or a group.

Typical applications, e.g. in companies, in associations, can be e.g:

+   Structured employee appraisals with active, mutual support,
+   Voluntary commitments with the active involvement of other stakeholders,
+   membership acquisition that strengthens the association,
+   the establishment and expansion of recurring neighborhood celebrations,
+   ... basically all issues that focus on integrating other interests, for example.

For some, a half-filled glass is half full when viewed positively, for others a half-filled glass is half empty when viewed negatively, despite the identical starting position.
Your perception influences your decision and if you surround yourself with negative people, you will not have any positive experiences.

So flip the switch, get out of the negative developments and you can do this by looking at the Johari window with the action-reaction relationship.

With a view to the energy transition, the Johari Window shows that all energy paths, from energy generation to applications, must be shown in full so that the energy transition can be recognized as important and correct.

The pro-fossil fuel lobby has long been unable to withstand a complete view of the energy transition, which is why the pro-fossil fuel lobby likes to communicate with deliberately incomplete statements, even deliberately false statements.

The extended view beyond the Johari window has the potential to identify all issues and promote community and/or democratic interests.

It is not always only manipulative influences from outside that make people look away from what is important and/or what is right, e.g. their own comfort can also be a trigger for looking away from what is important and/or what is right. The rebound effect, already shown on pages 14 and 15, is a suitable example of this.

Questioning is the most important basis for not falling for the numerous fake news and fake videos on the internet.

With the energy transition, very complex interrelationships are being optimized so that the optimum is not possible in all places at the same time; numerous developments precede and other developments follow.
At this point, I would like to ask once again, because there is a very clear need for gradual adjustments:

Does anyone really believe that the incandescent lamp developed by Thomas Alva Edison is the result right from the first approach?

The steam engine was given a decisive improvement in 1769 by James Watt, who was granted a patent for it.

Not everything that proves to be both sustainable and successful is applauded from the outset.

This is very true when it comes to the energy transition, partly due to uncertainty and partly due to fake news and fake videos circulating on the internet.

Before hydrogen entered the energy dialog, energy-producing companies and the operators of coal-fired power plants welcomed wind and solar energy as a green supplement, because energy production from wind and solar is not steady, but rather fluctuating.

As a result, wind and solar energy generation could not provide a base load and renewable energy generation could not form the basis for security of supply. This has changed with the addition of hydrogen; in times of excessive wind and solar energy generation, the excess electrical energy that the electrical grids cannot absorb can be used to produce hydrogen, which can be reconverted into electricity to make up for the shortfall in electrical energy in times when there is not enough wind and solar energy generation. Alternatively, with an eye on efficiency, the hydrogen is used directly.

From this point on, the pro-coal lobby has been fighting hydrogen with fake news and fake videos.

Voices were raised in favor of fossil fuels against hydrogen and a heated debate ensued; often with unjustified questions from the pro-coal lobby with the aim of slowing down the energy transition.

It is not uncommon for the loudest critics with a negative position to be rather nervous on closer inspection because they want to slow down the intended target.
In 1995, we are in the Emscher-Lippe region:
+   10 towns in the district of Recklinghausen,
+   Bottrop,
+   Gelsenkirchen
with the focus on hydrogen energy storage.

For the polite few we were visionaries, for the others we were "crackpots". In 2003, our sustainable results convinced the entire state and we were awarded the NRW state contract to coordinate hydrogen and fuel cell activities in the northern Ruhr region. Today we know that those who intervened against us with incomplete and deliberately false lobbying are themselves striving to enter the hydrogen market.

Confidence and clarity of purpose. It is not a question of interpretation, it is honest dialog that provides a strong basis.

Success is linked to the success of the dialog partners:
+   Identifying opportunities outside the routine,
+   combine curiosity with inspiration,
+   this also leads to unknown paths.

Discover the potential at the tangents and recognize the confirmation from the harshest critics. The more intense the motivation of the disruptors, the greater the potential grid with the resulting opportunities and possibilities.

<u>The energy transition is recognized today as important and right for environmental protection and for the maximum possible reduction of the effects of climate change.</u>

But this has not always been the case. In the beginning, the energy transition was seen as a disruption to the operational processes of energy generation.
Employees who spoke out in favor of the energy transition were perceived as disruptive and admonished, even threatened with transfer, for example.

There is a routine procedure before the disruption in the operational process. However, the disruption in the operational process opens up opportunities and possibilities that would otherwise often not even become apparent due to operational blindness.

Against this backdrop, fake news with the aim of slowing things down can arise.

The disruption in the operational process opens up several paths:
+   either the path that continues the action through the temporal exchange of the daily agenda,
+   or the path that gives free rein to spontaneity and creativity.

I prefer the second path, because it is the path with the antennae out, the one that combines curiosity with inspiration, the one with the desire for the unexpected, the one with the pleasure of variety.

<u>An alternative is not always crowned with success and those with a fleeting view tend to see the alternative as a mistake. However, mistakes are positive results in the first place.</u>

Results are the basis of all developments:

+	Please try to see a "mistake" as a decision-making aid.
+	I interpret as a real mistake that which must be prosecuted by the public prosecutor's office in accordance with the Criminal Code.
+	I interpret an antisocial behavior, such as egoism, as a mistake.

An idea or an opportunity that presents itself is sometimes judged by others as a "no".
Not so with me. The best way for me is that my "no" to others never stands alone.
The "no" becomes connecting and complementary and even rich in potential:

+	with "No, but don't we want ... instead?"
+	or with: "No, what do you think of the idea ...?",
+	or with: "No, together we have the chance ... instead".

Anyone who just wants to load me up with an idea will, once involved, not come back so quickly.
Anyone who actually seeks my involvement will receive a fair, honest and sustainable basis from me; my complementary design is both the driving force and the energy.

"Stay the way you are" is perhaps a wish expressed at work, in the family environment or in the outside world, and at first sounds unifying. However, "stay the way you are" is almost always an eloquent tool for the convenience of others, so that you conform to their model of life.

Our successes reflect our actions and train our ability to contribute to others. If we only orient ourselves to what is presented to us as successful with tempting, simple-sounding solutions, then we very quickly succumb to fake news and fake videos.
Those who are successful in the long term consciously act as part of a team and live by qualities such as loyalty, a sense of duty, caring, justice and courage.
Those who are successful in the long term act away from routines with spontaneity and creativity.
The result: a clear conscience quickly leads to well-deserved serenity.

<u>Giving opportunity acquisition the attention it deserves!</u>

Yes, a differentiated, self-questioning position opens up a wonderful amount of potential and opportunities; questioning creates very exciting impulses.
In fact, with a good basis of trust, I don't have to check everything in my private life or in a professional context, otherwise I would have to swap the wheels on my car for cubes to see whether the wheels or the cubes have a better braking distance, with all the resulting consequences for driving comfort.

Questioning things at school sometimes prompted the response from my teachers at the time: "Because that's the case, just read the *** book." From then on, it was clear to me that I would also write a book myself and that it would contain things that would enable others to question, that would invite others to join in the joint reflection and the acquisition of opportunities; even if I didn't have today's vocabulary back then as a child.

I have been creating reports since I was 21 years old. Yes, the terms commitment, trust and a sense of responsibility always come up when it comes to the success of cooperation in society and/or a career, for example. Life experience shows that people can still be problematic for others, despite their best references, due to their different points of view.

Honesty, even if it is uncomfortable for me, for example because I have forgotten something that I had previously agreed to; that is the fundamental basis. Honest dealings provide the important anchors of trust for finding solutions with inviting commitments.
Very responsible diplomacy towards others can strengthen the basis.
Does that sound far-fetched to you? Some people may be inclined to take this view, but in reality it is precisely these relationships that trigger valuable and enabling impulses.

<u>How do we want to make the important energy transition possible without the trust that binds us together?</u>

We know many things, but we need to work together to achieve something complete.

Yes, by stepping out of the routine and questioning:
+	you become visible to others,
+	both at work,
+	as well as in your private life
+	and you may initially become a little strange and vulnerable to those who are comfortable with you.

We all receive countless fake news and fake videos via social media, such as WhatsApp, which may even seem funny at first, but which are very often deliberately created to slow down the energy transition, for example, by creating the impression: If so many people are sharing and forwarding it, then there must be something to it.

Yes, questioning makes us visible and if we then counter the fake news and fake videos with our own, correct view, then we can become vulnerable.
That's why quite a few people remain silent when fake news and fake videos about the energy transition appear, giving the false impression that the fake message is widely accepted. Questioning takes courage.

<u>How do you proceed when you are allowed to decide?</u>
<u>Isn't that a very interesting question?</u>
<u>In fact, it's the all-important question!</u>

In all my keynote speeches over the years, I have noticed that there is definitely a difference:

+   between the generations,
+   which is also characterized by convenience and the fast pace of life.
+   While seniors actively participate in the dialog of my keynote speeches by questioning,
+   the same applies to the participants in the age group currently aged 40 and over,
+   the current age group of students, trainees and young workers is rather sluggish and I am increasingly recognizing that young people rarely question things, immediately jump on the current trend and live with the mainstream.
+   Where does that come from?

<u>Not everything that is published on Wikipedia, for example, is correct in terms of content. What is published on Facebook, for example, is often fake; nevertheless, the content is accepted and liked without being checked or even instrumentalized without being checked.</u>

Fake messages and forwarding, e.g. via WhatsApp, is omnipresent.

Do you remember the old saying: "What little Fritzi doesn't learn as a child, Fritz learns twice as hard as an adult"?

If people increasingly lose the ability to make decisions due to a lack of attention and a lack of willingness to question, triggered by convenience and the fast pace of life, the door is open to extremists, whether religiously or politically motivated, for example.

When I see today:
+ young parents pushing the baby carriage,
+ the stroller in one hand, the cell phone in the other,
+ when the child contacts the young parents by looking at them and/or talking to them and the young parents don't notice,
+ must we then assume that these young parents have not been sufficiently taught how to communicate with each other?
+ Perhaps. But how can this undesirable development be counteracted?

When you look at some TV shows, do you also realize that communication itself is incomplete and that this can cause problems? These TV shows are intended to amuse viewers with the reaction: "Fortunately, I'm not like that."

But what then can/should give young people valuable impulses and/or motivation?

A few years ago, a very interesting phenomenon called influencers emerged on social media.

Profession and/or vocation? The difference could not be clearer and more recognizable than with influencers, because in the end the result is also shaped by trust.

Trust, there it is again. The topic of trust is recognizably our important basis for working together, whether at work or in our private lives.

The conceptual explanation of influencer means that the English term "to influence" translates into German as take a different path.

This perhaps simple view does not do justice to the very high standards, because influencers are not just people who spread content in various media through text, images, audio and/or videos on their own initiative.

Influencers reach out on a topic that is characterized, for example: + by a trend (mainstream),
+   by a necessity (e.g. the energy transition in terms of environmental policy),
+   by an assignment (trade fair/congress/marketing) with the greatest possible frequency to reach a broad target group.

Quite a few young people aspire to a successful career as an influencer and fail due to a lack of persuasiveness, which can be due to:
+   a lack of experience,
+   insufficient knowledge of the tangents,
+   a lack of personal conviction.

Politics, sport, journalism, acting and social media are a basis for influencers, but only a few influencers have a lot of attention, a lot of followers. Why is that?

The hurdle is people's trust.
Trust is generated by leading by example with a focus on the important, desired results.

But what makes them successful influencers?
Is it just trust?
No:
+   It's definitely also the courage to be different,
+   the courage to swim against the tide,
+   the courage to endure the possible headwind, the shitstorm.

<u>The headwind, the shitstorm, has been experienced for many years by the players in the potential grid of the energy transition and this is exactly what I have actually encountered.</u>
<u>So how do you successfully implement the energy transition with trust?</u>

At the beginning of the book, I told you that we, at that time just four people in the Emscher-Lippe region, who gave presentations on the subject of hydrogen as an energy source in 1995, were visionaries to the polite few and simply cranks to the others.

In 1998, the perseverance and the constructive presentation of real potentials generated a great deal of supra-regional interest and popularity. Even if the word influencer did not exist in the period before the millennium year 2000, we were and are influencers.

The fact that we have succeeded in turning the corner, that we are no longer perceived as cranks, but that a supra-regional interest has arisen from numerous companies and organizations, right up to the state of North Rhine-Westphalia, has a basis that creates trust and there it is again, the trust and the courage to question and counteract.

Yes, the energy transition costs money. But the results of the insurance companies show that the effects of increasing climate change will lead to an increase in damage events such as storms, floods, floods and heat fires and that the expected costs will be much higher; probably by a factor of ten or more.

<u>Responsible influencers who not only focus their work on a broad target group, but also motivate people to work TOGETHER, must also think outside the box to motivate team building:</u>
+ Successful influencers have the courage to show themselves openly,
+ successful influencers are the department heads of their own ideas,
+ successful influencers are the ringmaster of their own enthusiasm
+ successful influencers are the dance teachers of their own inner drive.

Curiosity combines with inspiration the desire for the unexpected, with the pleasure of diversity.

I am constantly rediscovering myself and the potential at my tangents. On the one hand, this creates energy and, on the other, it creates the important serenity so that I can open myself up to chilling out, for example.

Keeping brain, heart and humor at eye level.

This often sounds rather nebulous to those under strain, but in fact it is nothing more than fairness towards oneself, one's fellow human beings and one's clients.
It's like the radio, there's not just off or on, there are many different settings from quiet to loud.

With Henry Ford's statement in mind, I place the will between "you can or you can't", which determines the respective dynamic.

It is then the dynamic that decides:
+	whether one is credible and finds support with the tendency "You can."
+	or whether you experience rejection with the tendency "You can't".

I myself am the engine and I should also "step on the gas" when I start up, because otherwise, as a permanent parker, boredom and frustration are the result.

Frustration and boredom are also a basis for susceptibility to fake news and fake videos.

If someone tells you that they can explain the energy transition to you within a few minutes, then you should be extremely skeptical.

Thinking EU/federal/state-wide and acting locally is not a contradiction, but rather a dynamic energy policy.

This book is an invitation to join the energy transition dialog and the following chapter provides suggestions with a view to the complementary topics of environment, climate and energy.

Suggestions

With the books I have published, currently 31 books, you will find supplementary and very detailed information in the areas of energy, climate and the environment.
I publish the books with BoD Verlag and all books are available from all book publishers and also from online marketplaces such as Amazon.

Books:
Left: DinA5, 88 Pages, The energy turnaround: Perhaps seeming paradoxical at first? Current positions.
Right: DinA5, 50 Pages, The siren-like songs and the fairy tales of full electrification. The sustainable energy transition links the electricity sector with heat, gas products and fuels. Hydrogen is the link with Power-to-X.

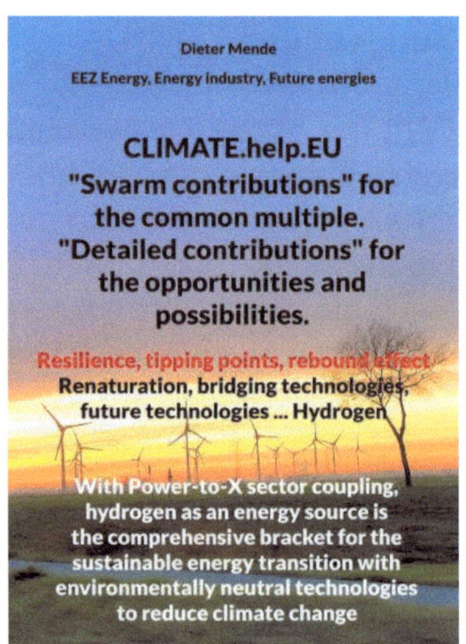

Books:
Left: DinA5, 60 Pages, The job engine of the energy transition. The global starting position.
Right: DinA4, 76 Pages, CLIMATE.help.EU : "Swarm contributions" for the common multiple. "Detailed contributions" for the opportunities and possibilities.
Below: DinA5, 62 Pages, Energy Energy industry Future energies, the fast track to the energy transition with hydrogen.

Books:
Above: DinA4, 68 Pages, ENVIRONMENT.help.EU : "Swarm contributions" for the common multiple. "Detailed contributions" for the feel-good oasis. A leafy arbor instead of a pavillon, imaginatively plantes small flowering islands for birds and insects, … Pond banks instead of „demolition craters" … use of recources …
Below: DinA5, 116 Pages, Patient Europe: a case for psychiatry? No! Release the handbrake, the hedgehog posture after Corona. People's reticence to the point of hysteria with conspiracy theories in the energy transition. The lobby, the intentions and the incentive are imbalanced.

The author

Dieter Mende

Homepage:
www.eez-mende.de

The book is written with the background of professional training in both chemistry and electrical engineering, supplemented with the background energy dialog EEZ Energie Energiewirtschaft Zukunftsenergien; I myself am the founder of the energy dialog EEZ on 05.07.1995.

My professional background is wide-ranging:
Commissioned since September 1997 with the central control technology for energy centers in the automation technology project office of a regional energy supply company; in November 2019, I moved to the electrical power grid construction project planning department.
Since February 2003, initially commissioned with the tasks of setting up and expanding the regional hydrogen nucleus h2herten, subsequently commissioned with tasks of project acquisition and company acquisition, market communication and networking in the team of the hydrogen user center h2herten.

In both cases, the employers noticed the very successful EEZ energy dialog, so that the jobs arose from this.

My motivation for creating reports and books is, on the one hand, a passion for highlighting the opportunities and possibilities in the potential grid of the energy transition with hydrogen as an energy carrier and, on the other hand, the ambition to build and expand a hydrogen infrastructure with the advertising of cross-industry service providers, with the identification of sustainable contributions and the resulting complementary competencies.

I met the established companies in the energy market and their initial rejection of hydrogen as an energy source and the fuel cell as an energy converter with meaningful results of the potential analyses, with the potential of the Power-to-X sector coupling, the identification of unique selling points and with the complex project impulses of the multi-layered interests of all those involved.
With perseverance and patience, even initial skeptics of hydrogen as an energy source and the fuel cell as an energy converter were successfully won over to a sustainable infrastructure in the energy transition.

With the EEZ Energy Dialogue, I am a long-standing member of the DWV German Hydrogen and Fuel Cell Association.

The DWV is one of the most successful mouthpieces in Europe for hydrogen as an energy source and for the fuel cell as an energy converter; the DWV speaks with the results of over one hundred industrial and research institutions.

I successfully contributed to the acquisition of members for the foundation of an advisory board for h2herten, from which the advisory board for the h2-netzwerk-ruhr emerged through expansion in 2008.

Picture above:
The hydrogen user center h2herten: the regional hydrogen nucleus in the northern Ruhr area.

Picture left:
14.06.2019 The opening of the hydrogen filling station at the h2herten user center; in the background the former coal mine Auf Ewald in the background.

Pictures: Dieter Mende; EEZ Energy Energy industry Future energies